SURVEY

Jim Crume P.L.S., M.S., CFedS

Co-Authors
Cindy Crume
Bridget Crume
Troy Ray R.L.S.
Mark Sandwick L.S.I.T.

PRINTED EDITION

*****

PUBLISHED BY:

Jim Crume P.L.S., M.S., CFedS

Bearings and Azimuths

Book 1 of this Math-Series

Copyright 2013 © by Jim Crume P.L.S., M.S., CFedS

All Rights Reserved

First publication: November 2013

Printed by CreateSpace

Available on Kindle and other devices

TERMS AND CONDITIONS

The content of the pages of this book is for your general information and use only. It is subject to change without notice.

Neither we nor any third parties provide any warranty or guarantee as to the accuracy, timeliness, performance, completeness or suitability of the information and materials found or offered in this book for any particular purpose. You acknowledge that such information and materials may contain inaccuracies or errors and we expressly exclude liability for any such inaccuracies or errors to the fullest extent permitted by law.

Your use of any information or materials in this book is entirely at your own risk, for which we shall not be liable. It shall be your own responsibility to ensure that any products, services or information available in this book meet your specific requirements.

This book may not be further reproduced or circulated in any form, including email. Any reproduction or editing by any means mechanical or electronic without the explicit written permission of Jim Crume is expressly prohibited.

Table of Contents

INTRODUCTION..4

BEARINGS...5

AZIMUTHS...6

EXPRESSING ANGULAR UNITS....................................7

DEGREES-MINUTES-SECONDS SYMBOLOGY....8

ADDING ANGLES TO BEARINGS...............................9

ANGLE BETWEEN TWO BEARINGS.......................14

DEGREES-MINUTES-SECONDS TO DECIMAL DEGREES...17

DECIMAL DEGREES TO DEGREES-MINUTES-SECONDS..20

BEARING TO AZIMUTH CONVERSION...............22

AZIMUTH TO BEARING CONVERSION...............22

ABOUT THE AUTHOR...24

INTRODUCTION
Straight forward Step-by-Step instructions.

This book is just one part in a series of digital and printed editions on Surveying Mathematics Made Simple. The subject matter in this book will utilize the methods and formulas that are covered in the books that precede it. If you have not read the preceding books, you are encouraged to review a copy before proceeding forward with this book.

For a list of books in this series, please visit:

http://www.cc4w.net/ebooks.html

Definition: Bearing or Azimuth. The lines or courses of boundaries, centerlines, etc. which describe the direction of a straight line between two points that are referenced to a meridian. The reference direction can be North or South and the meridian may be assumed, grid, magnetic, astronomic, or geodetic.

Note: True North is generally accepted as being astronomic or geodetic north. If a bearing meridian is not identified, it is assumed to be True North.

BEARINGS

A bearing angle is measured from the North axis, then left or right from 0° to 90° or from the South axis, then left or right from 0° to 90°. The bearing angle will be identified by the quadrant that it is being measured such as NE, SE, SW or NW. See **Figure 1** for the nomenclature of the Bearing system.

Note: A bearing will never be over 90°.

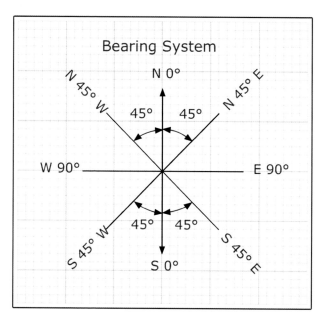

Figure 1

AZIMUTHS

An Azimuth angle is measured from the North or South axis in a clockwise direction from 0° to 360°. See **Figure 2** for the nomenclature of the Azimuth system.

Note: An azimuth will never be over 360°.

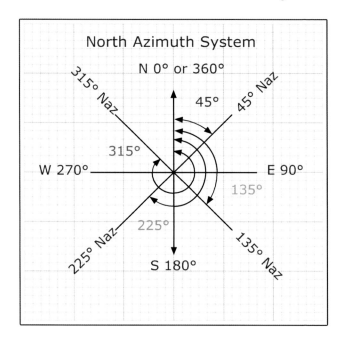

Figure 2

EXPRESSING ANGULAR UNITS

Angles are shown in Decimal Degrees, Degrees-Minutes-Seconds or Grads.

Note: For the purposes of this book, Decimal Degrees and Degrees-Minutes-Seconds will be utilized.

One degree equals 1/360 of a circle. In other words, 360 degrees equals a circle.

One minute equals 1/60 of a degree. In other words, 60 minutes equals one degree.

One second equals 1/60 of a minute. In other words 60 seconds equals one minute.

A circle equals 360 degrees or 21,600 minutes or 1,296,000 seconds.

This is known as Base 60 Arithmetic.

You can think of degrees-minutes-seconds being the same thing as hours-minutes-seconds for the measurement of time. The arithmetic is the same.

Several well known calculators have a button that is labeled Hours-Minutes-Seconds (->H.MS) & Hours (->H) such as:

This can also be interpreted as Degrees-Minutes Seconds (-> D.MS) & Degrees (-> D). More on this later.

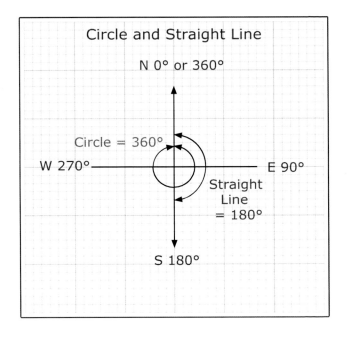

DEGREES-MINUTES-SECONDS SYMBOLOGY

Degree = º

Minute = '

Seconds = "

Example: North 20 degrees 45 minutes 10 seconds East can also be shown as N 20º45'10" E.

ADDING ANGLES TO BEARINGS

Example 1: Add 34°20'30" to N 20°45'10" E.

Line up the degrees, minutes and seconds for the bearing and angle in identical unit columns then add each column with the total at the bottom.

$$\begin{array}{r} N\ 20°45'10"\ E \\ +34°20'30" \\ \hline 54°65'40" \end{array}$$

Review the total for the seconds, in this example it is 40". It is not 60" or over so there is no reduction required.

Review the total for the minutes. The minutes value is 65' which is over 60' therefore it must be reduced to its degree and minutes equivalent of 1° 05' (65' - 60' = 5', 60' = 1°). The 1° will be added to the 54° making it 55°.

The final angle will be **55° 05' 40"**.

Now to determine which quadrant the resulting bearing is in. Since the final angle is less than 90°, it is in the same quadrant of NE as the original bearing therefore the resulting bearing is **N 55° 05'40" E.**

Note: For the final answer, all minutes and seconds will always be less than 60.

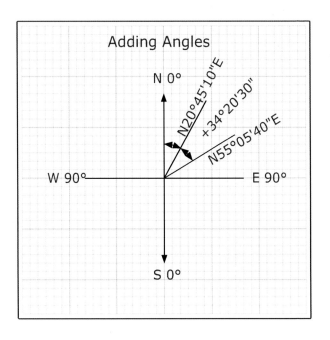

Example 2: Add 132°54'30" to N 35°20'35" W

Line up the degrees, minutes and seconds for the bearing and angle in identical unit columns then add each column with the total at the bottom.

$$\begin{array}{r} N\ 35°20'35"\ W \\ +132°54'30" \\ \hline 167°74'65" \end{array}$$

Review the total for the seconds. The seconds value is 65" which is over 60" therefore it must be reduced to its minute and seconds equivalent of 1' 05" (65" - 60" = 5", 60" = 1'). The 1' will be added to the 74' making it 75'.

Review the total for the minutes. The minutes value is 75' which is over 60' therefore it must be reduced to its degree and minutes equivalent of 1° 15' (75' - 60' = 15', 60' = 1°). The 1° will be added to the 167° making it 168°.

The final angle will be **168° 15' 05"**.

The final angle is over 90° and less than 180° therefore it must be subtracted from 180°.

Line up the degrees, minutes and seconds for 180° minus the final angle in identical unit columns then subtract each column with the remainder at the bottom. For the 180° value, you will have to borrow 1° or 60' for the minutes column then you will have to borrow 1' or 60" for the seconds column. 180° = 179°59'60"

$$179°59'60"$$
$$-168°15'05"$$
$$\overline{11°44'55"}$$

Now to determine which quadrant the resulting bearing is in. Since the final angle is greater than 90° and less than 180°, it is in the SW quadrant therefore the resulting bearing is **S 11°44'55" W**.

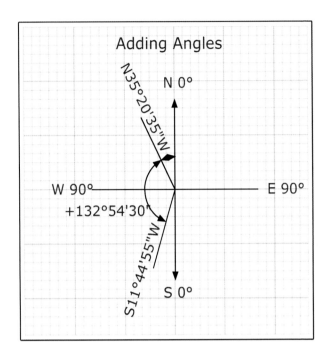

Note: It is helpful to draw a sketch, like above, to aid in determining the quadrant for the final bearing.

NOTES

ANGLE BETWEEN TWO BEARINGS

Known:

Bearing 1: N 34°12'23" W

Bearing 2: S 54°34'33" E

Solve: Calculate the clockwise angle between the two bearings.

Draw a sketch of the bearings to determine the steps needed to calculate the angle between them. When calculating angles between two bearings, often times it requires several steps to reach the final solution.

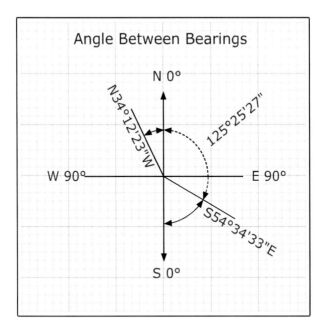

Bearing 1 is in the NW quadrant. The counter-clockwise angle from the north axis is 34°12'23".

Bearing 2 is in the SE quadrant. The counterclockwise angle from the south axis is 54°34'33".

Determine the angle for **Bearing 2** from the north axis by subtracting the angle of 54°34'33" from 180°.

$$\begin{array}{r} 179°59'60" \\ -54°34'33" \\ \hline 125°25'27" \end{array}$$

Now add the angle of 125°25'27" to 34°12'23".

$$\begin{array}{r} 125°25'27" \\ +34°12'23" \\ \hline 159°37'50" \end{array}$$

The resultant angle **159°37'50"** is the clockwise angle between the two bearings. Should you need the counterclockwise angle between the two bearings subtract 159°37'50" from 360° which equals **200°22'10"**.

Note: It is always helpful to look at each bearing separately and its relationship with the north or south axis. Once that is determined, then you can determine the best process to calculate the missing angle that will tie the two bearings together for the final answer.

NOTES

DEGREES-MINUTES-SECONDS TO DECIMAL DEGREES

You will want to utilize a calculator to add, subtract or perform trigonometric operations on angles. You will first need to convert Degrees-Minutes-Seconds to Decimal Degrees for ALL calculator operations.

Use the following steps to manually perform the conversions.

Formula: (((Seconds / 60) + Minutes) / 60) + Degrees = Decimal Degrees

---

Example: Convert the angle of 20°34'44" to decimal degrees.

(((44" / 60) + 34') / 60) + 20° = **20.57888889°**

Several calculators have a function that will perform this arithmetic.

For an HP calculator enter the angle as follows: 20.3444 then press the -> **H** button:

The calculator will parse the entry into its degrees, minutes and seconds components to perform the calculation.

The result will be 20.57888889 in decimal degrees.

Note: It is very helpful to have a calculator with the ->H function button that will perform the arithmetic, especially if you have many angles to add, subtract or determine the trigonometric values for.

NOTES

DECIMAL DEGREES TO DEGREES-MINUTES-SECONDS

Use the following process to convert from Decimal Degrees to Degrees-Minutes-Seconds.

D = Degrees
D.ddd = Decimal Degrees
M = Minutes
M.mmm = Decimal Minutes
S = Seconds
S.sss = Decimal Seconds

Formula:

D.ddd - D = .ddd
.ddd * 60 = M.mmm
M.mmm - M = .mmm
.mmm * 60 = S.sss [Round to nearest second as needed]

Final answer: D-M-S.sss or Degrees-Minutes-Seconds

Example: Convert 20.57888889° to Degrees-Minutes-Seconds

20.57888889 - 20 = 0.57888889 [20 is the degrees]
0.57888889 * 60 = 34.7333334
34.7333334 - 34 = 0.7333334 [34 is the minutes]
0.7333334 * 60 = 44.000004 [44 is the seconds]

Combine for the final answer: **20°34'44"**

Several calculators have a function that will perform this arithmetic.

For an HP calculator enter the angle as follows: 27.57888889 then press the -> **H.ms** button:

The calculator will solve the entry into its degrees, minutes and seconds components.

The result will be 20.3444 which reads 20 degrees 34 minutes 44 seconds.

Note: It is helpful to have a calculator with the ->**H.ms** function button that will perform the arithmetic, especially if you have a lot of decimal degree angles to convert.

BEARING TO AZIMUTH CONVERSION

1) For a NE bearing: North Azimuth = Bearing

2) For a SE bearing: North Azimuth = 180° - Bearing

3) For a SW bearing: North Azimuth = 180° + Bearing

4) For a NW bearing: North Azimuth = 360° - Bearing

AZIMUTH TO BEARING CONVERSION

1) For a North Azimuth of 0° to 90°: The NE Bearing = Azimuth

2) For a North Azimuth of 90° to 180°: The SE Bearing = 180° - Azimuth

3) For a North Azimuth of 180° to 270°: The SW Bearing = Azimuth - 180°

4) For a North Azimuth of 270° to 360°: The NW Bearing = 360° - Azimuth

NOTES

ABOUT THE AUTHOR
Jim Crume P.L.S., M.S., CFedS

My land surveying career began several decades ago while attending Albuquerque Technical Vocational Institute in New Mexico and has traversed many states such as Alaska, Arizona, Utah and Wyoming. I am a Professional Land Surveyor in Arizona, Utah and Wyoming. I am an appointed United States Mineral Surveyor and a Bureau of Land Management (BLM) Certified Federal Surveyor. I have many years of computer programming experience related to surveying.

This book is dedicated to the many individuals that have helped shape my career. Especially my wife Cindy. She has been my biggest supporter. She has been my instrument person, accountant, advisor and my best friend. Without her, I would not be the professional I am today. Cindy, thank you very much.

Other titles by this author:

http://www.cc4w.net/ebooks.html

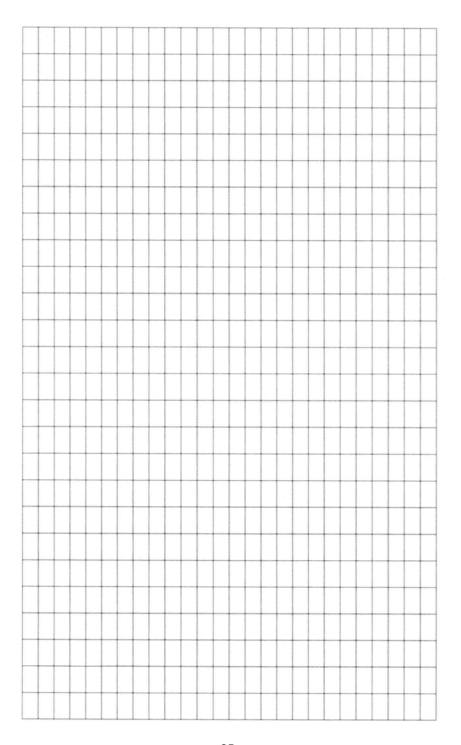

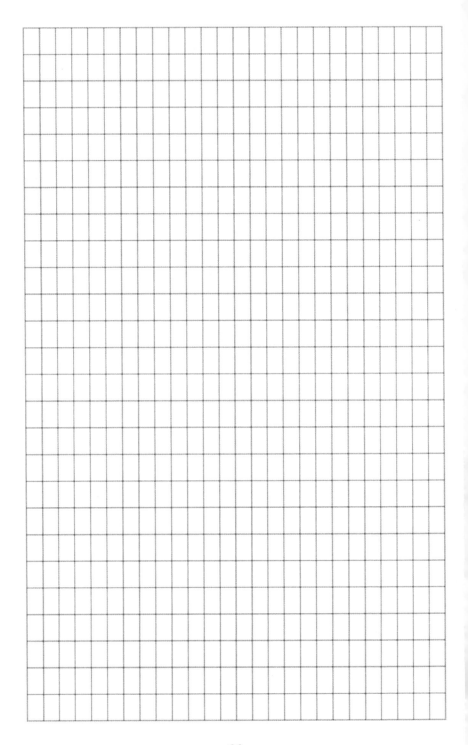

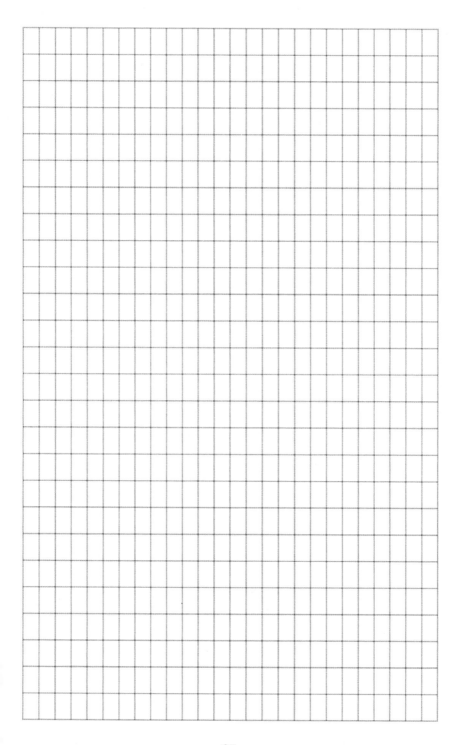

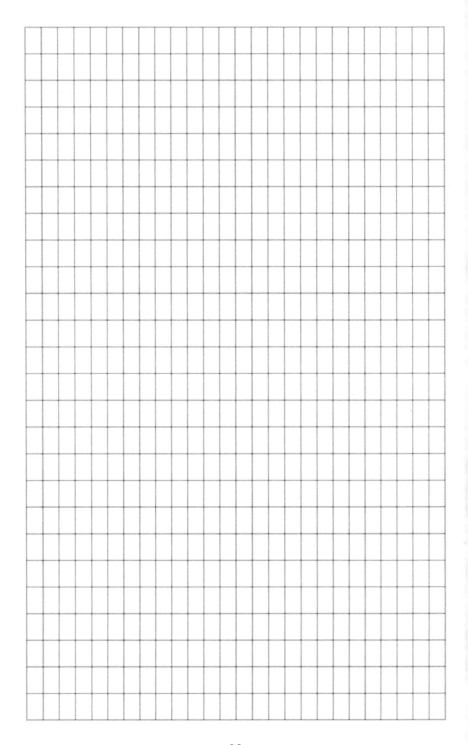

Made in the USA
Columbia, SC
26 February 2020